Acute Care –
Homeopathy for
Muscle Injuries

By D C Rona, PhD, ND, DHM

Acute Care - Homeopathy for Muscle Injuries

ISBN: 978-0-557-10562-5

Published by Lulu.com

Cover Photograph by Jessica Rona
Edited by Veneda B. Cook

Disclaimer

Forward

Alternative, integrative, complementary – all words used to describe healing modalities outside the traditional 'modern' western medical framework. Many of these modalities have been used and proven effective for centuries, and some for thousands of years.

Most often, holistic healing methods are centered on individualization – where 'modern' medicine is based on diagnosis and protocols of treatment of a diagnosis and not an individual. Herein lies the great schism - the reason it is so often easy to dismiss the holistic modalities. It is far more difficult to statistically 'prove' the effectiveness of individualization than standardization. Statistics and standardization are perhaps a simpler approach to medicine, but not a substitute for holistic healing.

For 200 years homeopathy has offered effective and safe results. It is fundamentally based on the body's inner abilities to heal and restore. True homeopathic healing results when the entire mind, body and spirit are considered in selecting remedies and potencies. However, it is also an exceptionally effective aide in acute situations – allowing the body to respond efficiently and allowing the whole system to work together to bring about a full recovery.

In the Acute Care series, the concepts and experience of applying homeopathy to a variety of acute situations is presented. Acute care is no substitute for the careful case taking and analysis of complete professional homeopathic care.

Homeopathy is completely complementary to traditional western modern medical care, and ideally should be integrated - using the best of each - for the care, healing and long term health of the individual.

Table of Contents

Chapter 1

What is homeopathy?

Throughout the history of mankind there have been attempts to understand the process of healing. The Law of Similars (like cures like) is found recorded in many places. A few examples include:

∞

"Through the like, disease is produced and through the application of the like is it cured."
Hippocrates (4th century BC)

∞

"You there bring together the same anatomy of the herbs and the same anatomy of the illness into one order. This simile gives you an understanding of the way in which you shall heal."
Paracelsus (5th Century AD)

∞

Samuel Hahnemann (1755-1843) was the founder of the system we call Homeopathy. Rediscovering the Law of similar, which he phrased as *Similia Similibus Curentur,* which translates as *"let likes cure likes"*, he developed a system using infinitesimally small doses to stimulate the organism to heal itself. He termed this medical system *Homeopathy* from the Greek words *homoios* for "similar" and *pathos* for "suffering" or "disease". Hahnemann's system is based on the principle which states that any substance, which can cause specific symptoms when given to a group of healthy people, can help to heal those who are ill and experiencing similar symptoms. The proper medicine, in very small amounts, promotes the reaction of healing within the body. In fact, he found that the smaller the dose, the more positive the cure.

> *Homeopathy is*
> *safe,*
> *gentle,*
> *and effective*

In 1810 Hahnemann wrote "Organon of Medicine". In subsequent revisions, he continued to refine and develop the system of homeopathy. Homeopathy became 'mainstream' medicine in the United States in the early 1800's. Numerous medical schools and hospitals still bear the name of Hahnemann. The first national medical association in the United States was the American Institute of Homeopathy, (founded in 1844). Politics (and not efficacy) prevailed in favor of pharmaceuticals and surgeries as the more 'modern' approach to medicine through most of the 20th century.

Because homeopathy works to support the vital force of the body, it assists the natural ability of the body to heal faster and does not suppress symptoms or delay the healing activities of the body. This is in contrast to pharmaceuticals that work by suppressing symptoms, even the symptoms that are healing responses (for example, acetaminophen suppressing a fever). Pharmaceuticals acting in this way can actually cause the body to take longer to recover from an acute injury (such as a sprain) or an infectious illness (such as cold or flu). Also, homeopathic remedies do not produce side effects (i.e. drowsiness) and they may be safely taken in conjunction with conventional medicines or treatments when necessary.

> *Homeopathy is the second most popular therapy in the world with over 300 million patients worldwide in 65 countries according to the World Health Organization (WHO).*

With an unrivaled safety record over the past 200 years, and a record of true effectiveness, it is easy to see why homeopathy is considered a mainstream medical therapy in Europe.

The concept of an individual's "constitution" is common to many systems of healthcare, including Homeopathy, Traditional Chinese, Indian Ayurvedic, and European Naturopathic. It holds that the individual has certain trends that are part of the basic nature of that individual. While one may thrive in open air and crave fatty foods, another feels best in cold and craves solitude and sweets, while another feels oppressed by open environments and seeks warmth. Of course, these examples only touch at the very tip of the entirety, but they serve to illustrate the concept. Each of these systems of medicine has a procedure for denoting and clustering physical, mental, emotional, and spiritual traits into constitutional patterns. By perceiving these patterns, one may be more effective in finding remedies that best fit that individual.

In homeopathy, the individual's constitution is identified after a lengthy interview process (and sometimes amended over time as the information is provided). This information is used to refine the search for the specific remedy that best fits the entire picture of the individual. Most often this information is used in treating long term disorders, but it may be incorporated into the choice of remedy for acute illnesses and even treatment after an accident. This type of investigation, analysis and remedy selection is part of 'classical' approach, adhering strictly to the original principles laid down by Hahnemann. By contrast, there are many versions of prescribing homeopathic remedies varying from using electrical devices to select remedies, to patterns of protocol for remedies by disease, even injecting solutions containing homeopathic materials – all of which are non-classical and non-Hahnemannian.

In working with a homeopath it's important to pay attention to all the reactions of the body. In relating information on any symptom, such as an aching back or sore throat, as, it is important to note specific details, such as:
- When did the symptom begin – and what were the circumstances
- An accurate description of the type of sensation (i.e. stabbing pain as opposed to burning pain)
- The exact localization (i.e. did the pain begin on the right side and move to the left, bore inward, etc.)
- The extension or movement of the pain (does it radiate up the foot, down the back, etc.)

- Does the symptom reoccur at a set interval of time?
- What makes the symptom better or worse (i.e. cold, heat, water, etc.)
- Is the symptom accompanied by any psychological change (i.e. sad, giddy, melancholy, etc.)
- Since the onset, has there been a craving for food or water (i.e. salt, sugar, cold water, etc.)

These are the details that will allow the homeopath to identify the correct remedy for the individual.

Homeopathy provides safe, natural effective remedies and individualized therapy, which encourages a cooperative relationship between patient and doctor.

Complementing mainstream 'traditional' research studies, two centuries of practice have shown the effectiveness of homeopathic medicines in treating a variety of musculoskeletal conditions including trauma, joint and muscle pain, fractures and sprains, and sciatica and lumbar pains. Depending on your activities and lifestyle, you may (over a lifetime) experience sport related injuries, home injuries, automobile accidents, or even industrial accidents. The basic application of homeopathic principles and remedies will be the same. Choice of homeopathic medicine as well as the frequency of dosage and the potency vary according to the seriousness of the injury and its acuteness. Old injuries which have become chronic will be treated differently than recent ones. The beauty of these medicines is in the value of the deep tissue, bone and nerve healing as well as pain relief. The homeopathic medicines used for acute injuries should be used in conjunction with conventional diagnostic tests (x-rays) treatment methods (splinting, casting, etc.) as well as basic care such as rest, ice packs, compression and elevation of the affected area.

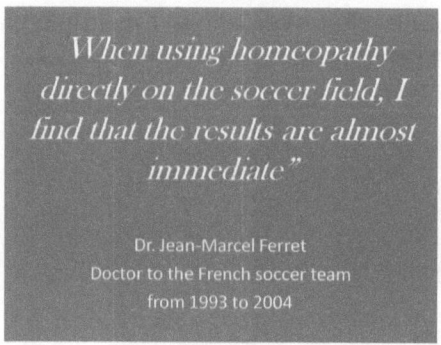

When using homeopathy directly on the soccer field, I find that the results are almost immediate"

Dr. Jean-Marcel Ferret
Doctor to the French soccer team
from 1993 to 2004

For those individuals who might be prone to frequent acute pain episodes – you may wish to work with a homeopath to select homeopathic remedies for immediate 'first aid' support until you can reach appropriate medical care. A skilled homeopath can help you select remedies that best fit your individual symptoms.

Homeopathic practices

Homeopathy is practiced as a complementary and/or alternative form of medicine. Homeopathy is frequently incorporated into the daily practice of other healing modalities including chiropractic, naturopathic, and sports physical therapist. In Europe, it is common to see homeopathy included in traditional western medical practices as an important modality. Acute application of homeopathic remedies requires minimum training and experience to achieve excellent results. However since several homeopathic medicines can address the same type of condition, anything beyond acute/emergency level treatment must take into account all of the particular symptoms of the individual. Appropriate education and training is necessary to utilize homeopathic remedies to their full potential. To work with any chronic condition requires full individualization, properly taking the case (history as well as details of symptoms and sensations) and analyzing not only the proper remedy but also the proper dosing. There are many schools of homeopathy throughout the world. For a listing in the United States and Canada, visit the website for the National Center for Homeopathy:

http://nationalcenterforhomeopathy.org

for schools, practitioners and other resources.

The jargon of homeopathy

Aggravation – the worsening of a symptom. This may be temporary homeopathic aggravation as the individual is healing, or disease aggravation if the underlying disease state worsens.

Constitution – the sum of the patient's symptoms, appearance, temperament, etc. that make it possible to select a remedy that fits their totality

Homeopathy – from the Greek homoios (like) and pathos (suffering) Homeopathy is a branch of medical therapeutics where the prescription of remedies is derived from the pharmacological principle of analogy (like cures like) and applied with small or infinitesimal doses. Samuel Hahnemann described this principle by using a Latin phrase Similia Similibus Curentur, which translates as "let likes cure likes." While known for centuries, Hahnemann applied this principle in developing a system of medicine called homeopathy. It has been used successfully for the last 200 years.

Materia Medica – a listing of the remedies with their corresponding symptom pictures both from proving and clinical experience

Modality – the factors which influence a symptom. These are the factors that make the symptoms worse or better for that individual, or the way in which a symptom moves

Polychrest – term given to 30-45 remedies that are used very frequently because they have such a wide range of applications

Potency – the strength of a remedy determined by dilution and succussion and denoted by X, C, M or LM depending on the dilution ratio used in preparation. The higher the dilution the 'stronger' or more potent the remedy.

Proving – experiments carried out by healthy volunteers to determine the symptom picture produced by each remedy

Remedy – homeopathic remedy – homeopathic medicine are terms used interchangeably in the discussion of homeopathic treatment and are used to describe a diluted and succussed medicinal substance which is 'proven' and included in the Homeopathic Pharmacopia.

Repertory – a cross reference system of indexing the symptoms (rubrics) produced by remedies

Vital force – the energy responsible for the state of health of an individual

Chapter 2

What constitutes an acute?

Dictionary definitions are simple black and white definitions. The reality of human experience is not so simple. Certainly there is a wide spectrum of situations. The car accident is the most simplistic illustrations of a pure acute situation – where the previous state of health or underlying past injuries did not contribute to the situation. A birth defect limiting kidney function in an adult is a simplistic illustration of a pure chronic situation. Most acute symptoms fall somewhere in between. The intense pain of a weekend athlete who's years of past injuries and poor training regime produce 3

or 4 episodes of knee pain per year could be considered an acute phase of a chronic problem. Likewise, severe

---

Acute
(from the Latin acutus, past participle of acuere – to sharpen, from acus – needle)

- Characterized by sharpness or severity (acute pain)
- Having sudden onset, sharp rise, and short course (acute disease)
- Being, providing, or requiring short term medical care (as for serious illness or traumatic injury)
- Lasting a short time

Chronic
(from the French chronique, from Greek chronikos of time, from chromos)

- Marked by long duration or frequent recurrence
- Always present or encountered, constantly vexing, weakening or troubling
- Being habitual

First Aid

Emergency care or treatment given to an ill or injured person before regular medical aid can be obtained

---

muscle cramps could be an acute phase of a chronic disease.

Within homeopathy, it is common to hear a general classification of acute vs. chronic diseases in discussing illnesses. Acute disease is usually defined as self-limited and brief, generally running their course within a reasonable period of time (i.e. colds, flu or viruses). Chronic disease is the term used for the situation where in individual suffers from repeated attacks of the same illness or the body presents a building set of symptoms. Acute diseases are often self treated or require only limited training to handle very well with homeopathy. Any time the situation is judged to be chronic it requires study and diagnosis by a homeopathic physician to be fully resolved (i.e. asthma, migraines and stomach disorders) without concern for suppression of symptoms.

Why is this important? Certainly homeopathic remedies may be applied to any acute situation, even those that are acute phases of chronic problems. However, if a chronic problem is the root of a current acute symptom – the individual will be best treated by fully individualized homeopathic treatment.

For the purpose of this book, we will focus on the use of homeopathic remedies for the typical acute situations that are seen in the course of normal day to day life - the intense pain and  limited mobility that accompany muscle injuries. Also, for the purpose of this book, we will focus on first aid assistance, assuming that appropriate, prompt and best possible medical diagnosis and treatment will be pursued or is taking place concurrently with the use of homeopathic remedies.

Homeopathic remedies are fast to act when used in acute situations and are helpful in pain relief and promoting the body's best healing responses. When homeopathic remedies are used in conjunction with conventional first aid procedures, the risk of long-term damage from an injury can be significantly reduced and the healing process is noticeably enhanced.

Homeopathy's greatest advantage as a component of acute "first aid" care is that it has no negative side-effects. It will not interfere with necessary traditional medical treatments and will enhance the body's ability to withstand the stress of procedures (such as the setting of a bone) and will usually significantly shorten the recovery time.

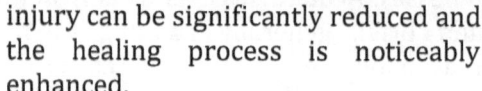

It is not uncommon to hear treatment and recovery defined in terms of phases or stages. Within different medical disciplines this breakdown will vary. In accidents, it is often broken into Phase 1 – the first 24 hours, and Phase 2 – the time after phase 1 in which the original symptoms persist. These descriptions are

arbitrary, and while they may be communication tools, they should not define a treatment protocol – since every individual must be evaluated as an individual and not some standardized protocol.

In general, in acute situations, well chosen homeopathic remedies should bring a sense of relief within a matter of minutes, with substantial relief felt within a matter of hours. The ability to sense relief depends on the individual, their current state of overwhelm in pain, their perception of fear over the situation, etc. The remedy may need to be repeated several times over the short term to achieve and hold the sense of relief. Of course, true healing takes time. The benefit of homeopathic remedies is that they not only have provided comfort and relief in the acute situation, but also they are supporting rather than interfering with the body's natural healing process, so recovery may begin immediately. Returning to health and vigor is of course the primary objective.

Chapter 3

Homeopathic remedies

Unlike some of the products found in health food stores - homeopathic medicines are made according to strict pharmaceutical and regulatory requirements. There are relatively few manufacturers of homeopathic remedies. A multidisciplinary team of pharmacists, chemists, zoologists and botanists must monitor every stage of homeopathic medicine manufacture.

The raw materials may come from animal, vegetable or mineral sources. They are screened and used in compliance with current pharmaceutical regulations, homeopathic tradition and official Pharmacopeia.

While the Europeans and Americans have slightly different 'standards books', any company preparing homeopathic remedies for sale in Europe or the United States must meet stringent requirements. The Homeopathic Pharmacopoeia Convention of the United States is responsible for establishing good manufacturing procedures and industry standards in collaboration with the FDA. In the US, the Food and Drug Administration (FDA) has recognized the Homeopathic Pharmacopeia of the United States (HPUS) since 1983. As an official reference on the subject, it enables homeopathic stocks to be registered. In order  to be placed on the American market, the homeopathic medicine must meet the conditions established by the FDA in 1998.

These homeopathic medicines are most often found on the store shelves packaged in granule tubes or globule doses. The vocabulary of the pill sizes varies widely depending on the training of the individual or the country of origin of the medicines. In homeopathic references you will see the tiny pills described as pills, pillules, granules, globes and tablets. There are also other forms such as pomades, gels, syrups, caplets, eye washes, drops, suppositories, etc.

Remedy names are given in Latin, which provide the most consistent means for accurately identifying the remedy. You may note that homeopathic names may differ from common names that are used when in herbal form. A good example of this is the homeopathic remedy Crataegus which is more commonly known as Hawthorn.

There will be abbreviations used on labels, simply because they run out of room. An example is Rhus Toxicodendron (poison ivy) which you will often see abbreviated as Rhus tox.

A common error in referring to homeopathic remedies comes when only part of a name is used. For example, you will frequently see reference to Aconite or Aconitum. However, Aconite is available in a variety of forms including:

Aconitum cammarum
Aconitum ferox
Aconitum lycotonum
Aconitum napellus

Normally it is Aconitum napellus that is available in local health food stores. The other varieties would most likely need to be special ordered. However, when looking at a remedy name in a reference book, such as a Material Medica or on a label – it is important to ensure you are looking at the full name of the remedy to insure accuracy.

The potency number is found on every remedy, immediately following or below the remedy name. Homeopathic medicine consists of natural substances that are extremely diluted. The amount of dilution is called potency. In addition to dilution, remedies are also 'succussed' in the process. Succussion is shaking with impact (the early preparations were made by holding the dilution bottle in the hand and slamming the bottle/hand onto a book), which has been shown to enhance the therapeutic effect of the remedy.

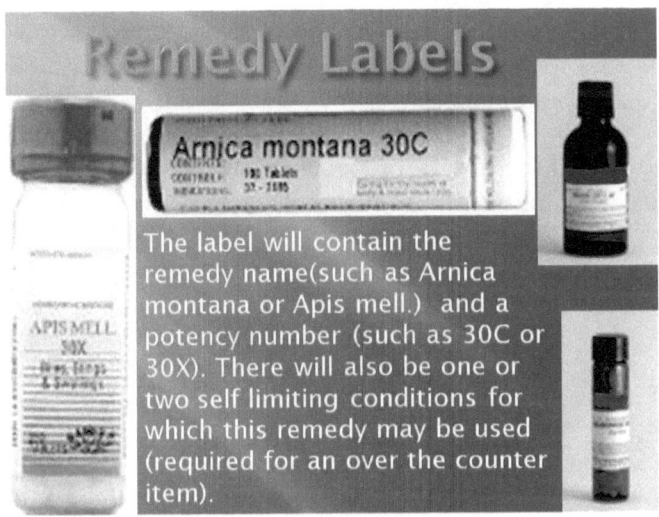

The "X" after the potency number (as in 6x) refers to the number of times in which a medicine is diluted 1:10, while the "C" after the potency number (as in 6c) is diluted 1:100. There is also an M scale which is an extension of the C scale. The higher the potency used, the more accurate the remedy selection must be. Also the higher the potency, the more often the psychological factors are involved and affected. Because of this, it is usually recommended to use only up to the 30th potency unless you are an experienced and well trained homeopath. Additionally, 200 years of homeopathic clinical experience has found that between the 6th and 30th potency are exceptionally effective for acute physicals, and they may be repeated more often. Some homeopaths contend that the lower potencies have a greater affinity for physical trauma. Emergency treatment most often utilizes 6x, 6c, 12x, 12c or 30c.

In his later years, Dr. Hahnemann developed the LM scale of potencies, using dilutions on the 1:1000 scale. These were by far his most potent and yet gentlest remedies. These are described in his final (6th) edition of the Organon of Medicine. Many of the most effective homeopaths in the intervening 200 years did not know of this last development since the 6th edition was not published immediately. Now, due to the exceptionally effective legacy of those 4th and 5th edition practitioners, and due to the nature of the dosing of the LM potencies, most remedies available in our neighborhood health food stores are X or C potencies.

> *Note: In Europe the D (decimal) scale is often utilized. A wide range of remedies is on hand at almost every neighborhood pharmacy, and often prescribers will use a greater variety of potencies than we find readily available in the US.*

Low potencies (i.e. 3x, 6x, 6c, 12c) are generally used for more localized issues and for acute situations. Higher potencies (30c and higher) are generally used when the patient's general health is a factor or where psychological influences are important to the case. These simple rules of thumb are just that – and many times an experienced

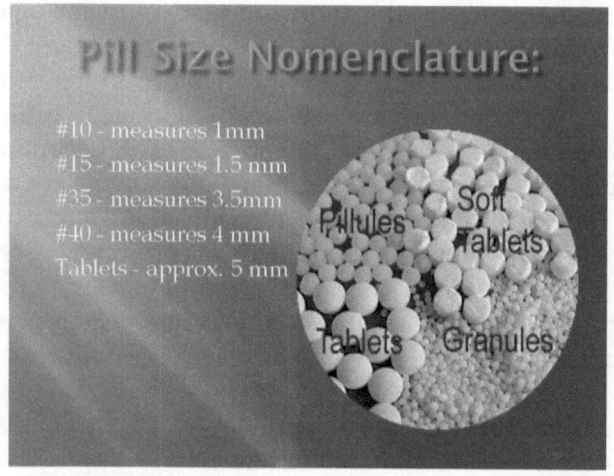

Pill Size Nomenclature:

#10 - measures 1mm
#15 - measures 1.5 mm
#35 - measures 3.5mm
#40 - measures 4 mm
Tablets - approx. 5 mm

Pillules
Soft Tablets
Tablets
Granules

practitioner will use potency based on additional factors such as the general health and energy level of the patient. Higher potencies are best prescribed by the more experienced homeopathic practitioners since the accuracy of the remedy choice becomes more critical. Low potencies are often repeated more frequently, as the patient feels it necessary, while higher frequencies are not repeated often if at all. In all use of homeopathy, the accurate selection of the remedy is far more important than the accurate selection of potency.

Single Remedies and Combination Formulas

Homeopathic medicines are available as single remedies or as formulas of two or more remedies mixed together. Herein lies great contention between those classically trained homeopaths and those who have come to use homeopathy as a 'this for that' treatment - which is an allopathic approach that simply uses the excellent effects of homeopathic remedies. The classically trained homeopath will stand firm that the ONLY way to treat homeopathically, using homeopathic remedies, is to individualize the single remedy that most clearly fits the presenting complaints of the individual. Anything other than complete individualization is not homeopathic – since at its very core, classical homeopathy is TOTALLY individualized. Others who have found homeopathic remedies to be exceptionally useful, but do not have the time or training to completely individualize, have found that it can be effective to mix a few remedies that are 'usually' applicable to a particular situation, and give that mixture to anyone in a specific situation. You will find even highly reputed homeopathic manufacturers that will put out a line of these combination remedies in addition to the single remedies since they are effective, and they meet the needs of a segment of the population. Then there are a few individuals who have pushed that boundary even further and put a mixture of tens of remedies into a combination, and still contend that they are using 'homeopathy'.

Single remedies are more effective for any situation when you determine the correct medicine. By giving a single remedy, you have the option to use the precise potency (or change the potency) to meet the needs of the individual, or change the remedy as the symptoms change. Combination or 'formula' products usually contain remedies in the 3, 6, or 12th potencies. Severe pain in otherwise healthy and vigorous individuals may respond more rapidly by selecting from 30 C or 200 C potencies. When combination remedies work, you do not know which component was working, so if you desire to continue treatment and need to change potencies, you do not know which single remedy to use. When combination remedies do not work, you have wasted precious healing time, and have no idea if it were the individual remedy components that were not effective, or if one worked well and the others neutralized its effect.

Remedy Information Resources

Homeopathic remedy information is indexed into two types of references; Materia Medica and Repertories. The Materia Medica is an organized description of each homeopathic remedy. Depending on the individual Materia Medica, you may find a quick short 'snapshot' of the remedy or you may find encyclopedic coverage. To be useful to a professional homeopath, the Materia Medica will give various characteristics and symptoms a person might exhibit who needs that specific remedy. Also commonly listed are clinical uses for the remedy, keynote symptoms, modalities and the full range of proving symptoms.

Modalities are listings of what makes someone feel better or feel worse. For instance, Bryonia will list the modality of *worse with motion*, whereas Rhus tox will list the modality of *worse with inactivity*. Correctly applying these modalities in the search for the best remedy for an individual is straightforward, but a learned and practiced art.

Homeopathic remedies go through a process of "proving" to determine the curative properties of each remedy. In a proving, health humans are given the item and the symptoms produced are carefully recorded. Homeopathy's most basic premise is Like Treating Like or Similar Treating Similar. So symptoms produced in a proving are the same symptoms that can be treated with that remedy.

The following example of a very short Materia Medica for Arnica is taken from *Boericke's Materia Medica*, 1901, written by William Boericke.

Arnica montana (*Leopard's Bane*)

Produces conditions upon the system quite similar to those resulting from injuries, falls, blows, contusions. Tinnitus aurium. *Putrid phenomena.* Septic conditions; prophylactic of pus infection. Apoplexy, red, full face. It is especially suited to cases when any injury, however remote, seems to have caused the present trouble. *After traumatic injuries,* overuse of any organ, strains. Arnica is disposed to cerebral congestion. Acts best in plethoric, feebly in debilitated with impoverished blood, cardiac dropsy with dyspnoea. A muscular tonic. Traumatism of grief, remorse or sudden realization of financial loss. Limbs and body ache as if beaten; joints as if sprained. Bed feels too hard. Marked effect on the blood. Affects the venous system inducing stasis. Echymosis and haemorrhages. Relaxed blood vessels, black and blue spots. Tendency to haemorrhage and low-fever states. Tendency to tissue degeneration, septic conditions, abscesses that do not mature. *Sore, lame, bruised feeling.* Neuralgias originating in disturbances of pneumo-gastric. Rheumatism of muscular and tendinous tissue, especially of back and shoulders. Aversion to tobacco. *Influenza.* Thrombosis. Hematocele.

Mind.--Fears touch, or the approach of anyone. Unconscious; when spoken to answers correctly, but relapses. Indifference; inability to perform continuous active work; morose, delirious. Nervous; cannot bear pain; whole body oversensitive. Says there is nothing the matter with him. Wants to be let alone. Agoraphobia (fear of space). After mental strain or shock.

Head.--*Hot, with cold body;* confused; sensitiveness of brain, with sharp, pinching pains. Scalp feels contracted. Cold spot on forehead. Chronic vertigo; objects whirl about especially when walking.

Eyes.--Diplopia from traumatism, muscular paralysis, retinal haemorrhage. Bruised, sore feeling in eyes after close work. Must keep eyes open. Dizzy on closing them. Feel tired and weary after sight-seeing, moving pictures, etc.

Ears.--Noises in ear caused by rush of blood to the head. Shooting in and around ears. Blood from ears. Dullness of hearing after concussion. Pain in cartilages of ears as

if   bruised.
Nose.··Bleeding after every fit of coughing, dark fluid blood.   Nose   feels   sore;   *cold.*
Mouth.··*Fetid breath.* Dry and thirsty. Bitter taste (*Colocy*). *Taste as from bad eggs.* Soreness of gums after teeth extraction (*Sepia*). Empyaema of maxillary sinus.
Face.··*Sunken;* very red. Heat in lips. Herpes in face.
Stomach.··Longing for vinegar. Distaste for milk and meat. Canine hunger. Vomiting of blood. Pain in stomach during eating. Repletion with loathing. Oppressive gases pass upward and downward. Pressure as from a stone. *Feeling as if stomach were passing against   spine.   Fetid   vomiting.*
Abdomen.··Stitches under false ribs. Distended; offensive flatus. Sharp thrusts through abdomen.
Stool.··*Straining of tenesmus in diarrhoea. Offensive,* brown, *bloody,* putrid, involuntary. Looks like brown yeast. Must lie down after every stool. Diarrhoea of consumption; worse lying on left side. Dysenteric stools with   muscular   pains.
Urine.··Retained from over·exertion. Dark brick·red sediment. Vesical tenesmus with very painful micturition.
Female.··Bruised parts after labor. Violent after·pains. Uterine haemorrhage from mechanical injury after coition. Sore nipples. Mastitis from injury. Feeling as if foetus   were   lying   crosswise.
Respiratory.··Coughs depending on cardiac lesion, paroxysmal, at night, during sleep, worse exercise. Acute tonsillitis, swelling of soft palate and uvula. Pneumonia; approaching paralysis. Hoarseness from overuse of voice. Raw, sore feeling in morning. Cough produced by weeping and lamenting. Dry, from tickling low down in trachea. Bloody expectoration. Dyspnoea with haemoptysis. All bones and cartilages of chest painful. *Violent spasmodic cough, with facial herpes.* Whooping cough, child cries before coughing. *Pleurodynia   (Ranunc;   Cimicif).*
Heart.··*Angina pectoris;* pain especially severe in elbow of left arm. Stitches in heart. Pulse feeble and irregular. Cardiac dropsy with distressing dyspnoea. Extremities distended, feel bruised and sore. Fatty heart and hypertrophy.
Extremities.··Gout. Great fear of being touched or approached. Pain in back and limbs, as if bruised or beaten. Sprained and dislocated feeling. Soreness after

overexertion. Everything on which he lies seems too hard. Deathly coldness of forearm. Cannot walk erect, on account of bruised pain in pelvic region. Rheumatism begins low down and works up (*Ledum*). **Skin.**--*Black and blue*. Itching, burning, eruption of small pimples. *Crops of small boils* (*Ichthyol; Silica*). Ecchymosis. Bed sores (*Bovinine locally*). Acne indurata, characterized by *symmetry in distribution*. **Sleep.**--Sleepless and restless when over tired. Comatose drowsiness; awakens with hot head; dreams of death, mutilated bodies, anxious and terrible. Horrors in the night. Involuntary stools during sleep. **Fever.**--Febrile symptoms closely related to typhoid. Shivering over whole body. Heat and redness of head, with coolness of rest of body. Internal heat; feet and hands cold. Nightly sour sweats. **Modalities.**--*Worse*, least touch; motion; rest; wine; damp cold. *Better*, lying down, or with head low. **Relationship.**--Antidotes: *Camph. Vitex trifolia.* Indian Arnica (Sprains and pains, headache in temples, pain in joints; pain in abdomen; pain in testicles). Complementary: *Acon; Ipec.* Compare: *Acon; Bapt; Bellis; Hamam; Rhus; Hyperic.*

Homeopathic vocabulary can sometimes appear confusing. The original works of Hahnemann and the early homeopaths were written in German and many of the early translations used a combination of terms common to the practice of medicine in that time period. Translations will often contain archaic sounding descriptions, but they are no less valuable. For example, feeling nauseous when riding in a carriage is completely equivalent to motion sickness from today's car, plane or train.

Repertories are a cross reference to remedies where the symptoms or illnesses are grouped by character and/or body part, and the remedies known to have those particular symptoms are listed.

The following short example of a repertory listing is from THE REPERTORY by Oscar E. Boericke, M.D.

From · LOCOMOTOR SYSTEM

SHOULDERS·SCAPUAE

Deltoid, pain, rheumatism ·· Ferr. p., Glycerin, Lycopers., Med., Nux m., Ox. ac., Rhus t., Sang., Sticta, Syph., Urt., Viola od., Zinc. ox., Zing.

PAINS ·· Alum., Am. phos., Anag., Arn., Azadir., Bar. c., Can. ind., Chel., Cocc., Con., Fagop., Jugl. r., Kreos., Lyc., Menisp., Myr., Nat. ars., Nat. c., Nit. ac., Ran. b., Sep., Ver. a., Viscum.

Scapulæ [between] ·· Æsc., Ars., Bry., Camph., Chenop., Euonym., Granat., Guaiac., Mag. s.

Burning, in small spots ·· Agar., Phos., Ran b., Sul.

Drawing ·· Ars., Berb. v., Cham., Col., Sul.
Left [in] ·· Acon., Æsc., Agar., Ant. t., Aspar., Cham., Chenop. gl., Col., Eup. purp., Ferr. m., Led., Lob. syph., Nux m., Onosm., Rhodium, Stram., Sul.

Lower angle ·· Chenop. gl., Cupr. ars., Ran. b.

Right [in] ·· Abies c., Am. m., Bry., Chel., Chenop. gl., Col., Ferr. p., Ferr. mur., Guaco, Ichthy., Ipomuea, Jugl. c., Kali c., Kal., Mag. c., Pall., Phyt., Puls. nut., Ran. b., Sang., Solan. lyc., Sticta, Stront., Urt.

Lower angle ·· Chel., Chenop. anth., Kali c., Merc., Pod.

Rheumatic ·· Acon., Am. caust., Berb. v., Bry., Colch., Ferr. mur., Ferr. p., Guaiac., Ham., Kali c., Kal., Lact. ac., Led., Lith. c., Lith. lact., Med., Ol. an., Pall., Phyt., Prim., Radium, Ran. b., Rhod., Rhus t., Sang., Stellar., Sticta, Stront., Sul., Syph., Urt., Viola od.

Stitches, tearing ·· Am. m., Bry., Hyper., Kali c., Lyc., Mag. c., Nit. ac.

Stiffness ·· Cocc., Dulc., Granat., Indium, Phyt., Prim. v., Sang., Senec. jac

Expiration date for homeopathic remedies

You will usually see an expiration date on homeopathic labels of approximately 5 years from date of manufacture. There is no reason (other than labeling laws) for this date since remedies made 200 years ago are still useful. The usual warnings to keep remedies in cool, dry places out of direct sunlight are wise for any medicine. Most also consider it wise to keep homeopathic remedies away from intense magnetic fields and intense substances such as mothballs, essential oils, etc.

Sources for homeopathic remedies

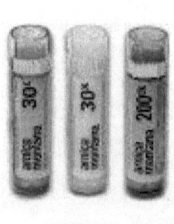

Many health food stores carry some homeopathic remedies – and most often they will carry all of the basic 'first aid' remedies in 12c or 30c. This is adequate for most situations. Most tubes will contain roughly 80 pillules (depending on the size of the pillules), easily fits in the pocket or purse, and will usually cost between $5 and $9 depending on the store. If you are constructing an emergency first aid kit for the office, a sport team field kit, hiking or travelers kit, or a home disaster kit you may wish to use a homeopathic pharmacy (either in person or online) to obtain the remedy selection and potencies that best match your needs. A professional homeopath can help you design a kit to match your specific needs. Then use a search engine for "Homeopathic Remedy Online Store" to get a current listing of online sources.

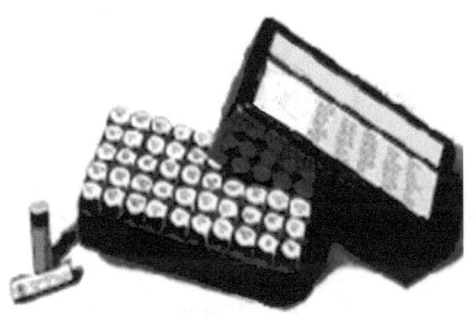

Chapter 4

When, which and how to use homeopathic remedies

Choosing the Remedy

Selecting the exact remedy to meet the full needs of an individual requires significant training and experience and an in depth interview to ascertain the full picture. Selecting an excellent first aid remedy to meet the needs of an acute situation is relatively easy and will significantly aid in

immediate comfort and long term healing of the acute. Each remedy has a profile of the physical and psychological aspects that it covers. These profiles are found in a Materia Medica and can be voluminous, truly encyclopedic detailed portraits. Each remedy also has a shortened set of 'key notes' which can give you a quick 'snap shot' of its uses. These key notes are something like an outline or short synopsis of a massive book. You might use such an outline to decide between War and Peace and The Iliad and Odyssey, but it would not provide anything of the true depth, color and quality of the books. Key notes are wonderful to use in emergencies, but any time you can, using the full Materia Medica will give a better understanding of the remedy potential. Keynotes may occasionally be misleading. For example, a keynote for a remedy may be 'lack of thirst' but the remedy may be exceptionally useful for an individual feeling great thirst on some occasions. Overall, you are very safe in using keynotes for first aid situations.

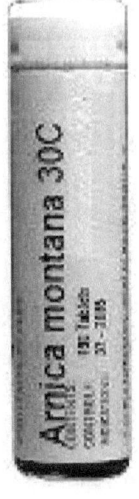

For acute treatment at home, or out hiking, biking, or coaching soccer, a small number of well documented acute remedies may be used with excellent results. In this case, you only need a short description of each remedy to make a good selection. You will find this sort of information in Section 5. Should you find these remedies do not appropriately cover your needs or the events you encounter most often, you may create your own quick reference by consulting with a professional homeopath.

Taking a Remedy

It is very convenient to keep the small tubes or bottles of remedies in a first aid kit. Usually to dispense from the small tubes you hold the tube upside down and twist the plastic top until 2 or 3 pillules are in the clear plastic top. You then pop the pillules directly under the tongue and let them dissolve. With the small bottles, you tip 2 or 3 pillules into the top or onto a clean spoon and again, pop them under your tongue to dissolve. This is a very effective dosing technique and is called 'dry dosing'. Larger bottles or jars of remedies require the pills or tablets be dispensed into a spoon, paper, or other such method prior to placing them in the individual's mouth – if they are to be taken dry. Professionals may dispense a single dose (2 to 4 pillules) placed in a neatly folded paper or small envelope with instructions to place the pillules under the tongue at a specified time.

When possible, the remedy should be placed in water (spring or distilled is preferred over tap water) and sipped. This is commonly referred to as 'wet dosing'. There is an entire methodology of dilution and 'wet' doses within homeopathic practice, but for acute care, simply dissolving the pills into water will greatly speed the action of the remedy. It also permits you to alter the potency between dosing by shaking or stirring the remedy solution. This minor change permits the best possible healing potential.

For ongoing treatment of an acute, it is common to prepare a small dropper bottle (30 ml or smaller) so the individual may continue to take doses through the healing process. To prepare such a bottle the instructions may call for 3 to 10 pillules into distilled or spring water. Drops are then dispensed directly into the mouth or into a glass of water at a prescribed timing, or as needed, with several intense shakes between doses. This dilution may also be applied to the temples or wrist in some situations. If the bottle is kept for more than a day or so it is appropriate to add a drop or two of vinegar or brandy to aid in preservation.

During competitions and challenging training sessions athletes sometimes add Arnica to the regular water bottle and sip throughout the competition or training session. In these situations, low potencies (under 30c) are usually selected. After a major trauma a professional homeopath might suggest higher potencies such as 200c or 1M be used for severe injuries and pain. This requires accurate remedy selection.

Avoid Contamination

Homeopathic remedies are remarkably stable and forgiving, but you can take steps to keep them at their most effective. Rough handling with dirty fingers can potentially contaminate the remedies and it is always advised not to directly touch or handle the dry remedy pillules – but rather take them from the dispensing cap, a clean spoon, paper, etc. If you spill the remedies, do not put them back in the bottle.

To keep remedies effective for years - avoid heat above 120 degrees, moisture, and extended periods of sunlight. It is best to store at room temperature in a drawer or box, away from chemicals and electromagnetic fields. Dry remedies should not be stored in the refrigerator; however wet remedies may be stored in the refrigerator.

Dosing strategies

We are used to standardized dosing protocols in the traditional pharmacological industry. One size fits all. In homeopathy, dosing instructions vary widely. An experienced prescriber will evaluate the situation, the sum of the individual's injuries, the overall 'vital energy' of the individual, and the events yet to be faced by the individual (i.e. medical procedures, wait times in the emergency room, etc.) before determining a dosing strategy, and even then, it will change with feedback from the injured individual. In most emergency situations, the dosing begins at a low potency (usually 6x, 6c, 12x, 12c, 30x or 30c) every 15 to 20 minutes for severe pain. As slight improvement is noticed, the frequency is reduced, usually to every 1 to 4 hours, or stopped altogether. The remedy is repeated then only as needed. Rarely do homeopathic remedies need to be repeated after the first few days. [Note: chronic conditions take longer.] The exception to this is a combination of physical re-injury (i.e. setting of a broken bone), chemical antidoting (i.e. pharmaceuticals), emotional aggravations (if you have been in a hospital recently you know what these are), and other forms of antidoting of the remedy (see the next section). Very rarely an individual may experience a homeopathic aggravation (an overstimulation of the healing process) after taking a remedy – in which case once the remedy is stopped, the aggravation will subside, and you may consider retaking the remedy – usually at a lower potency. Homeopathic aggravation is exceptionally rare in emergency situations with appropriate acute remedies.

Very rarely two remedies are well indicated for the situation. If that is the case, best results are found by waiting about 15 minutes between taking each remedy.

Switch remedies if the symptoms change and begin to show that another remedy better fits the situation. Discontinue all homeopathic remedies when improvement is stable and steady. Let the body do its job, and support with nutrition, water, rest, etc.

The most important factor of homeopathic treatment is the selection of the correct remedy. No matter what potency the remedy, if it is the right remedy, in an emergency situation, it will have a healing effect on the body. If taken properly, no side effects or harm should ever occur. Remember, it is seldom necessary to take a homeopathic remedy for a long period of time.

Antidoting

It is possible to neutralize the effect of a homeopathic remedy and render it useless. Remedies should be kept away from airborne contaminants such as perfumes, aromatic oils and camphor. Smoking and caffeine can affect the effectiveness of remedies. Mint is generally considered to neutralize or counteract the action of homeopathic remedies. It is wise to not use any stimulants for two hours before and after taking homeopathic remedies, if possible. The following could possibly antidote the remedy:

- coffee
- mints
- strong smelling linaments, e.g. tiger balm, vicks vapo rub, eucalyptus, or camphor,
- perfumes & colognes

In an emergency situation – just give the remedy and don't worry about anything else. If you find the remedy is not acting effectively, and you are confident it was the right choice, then look at the possible antidoting factors and re-dose.

External Applications

 While we most often think of homeopathic remedies as 'internal' remedies, there are several homeopathic remedies that are available for topical applications. Most often you will find Arnica, Calendula or Hypericum in creams, ointments, gels and even sprays. They are effective and easily used. Select the remedy based on the same information you would use to select internal remedies – and select the vehicle based on the specific situation. Ointments are usually placed in a petroleum base, which means they are not easily washed or wiped off, but do not allow the skin to breath. Gels and sprays allow the skin to breath but do not remain long on the skin of an active individual. Creams are usually comforting and soothing so are more often used at bedtime.

Several additional factors must be considered with external applications. Calendula is excellent for healing cuts and scrapes. Arnica is excellent for bumps and bruises, but should not be used with any open skin (i.e. cuts and scrapes). These specific directions will be found on the packaging for each product.

Suggestions for Taking Homeopathic Remedies
(under ideal situations)

- Take nothing by mouth 10-20 minutes prior to or following dosage. This includes food, drink, chewing gum, mouthwash, toothpaste etc.
- Limit as much as possible caffeine in any forms, such as soft drinks, coffee, or chocolate.
- No camphor, such as in muscle and joint rubs (e.g. sports balm, vapo rub, eucalyptus creams)
- Avoid moth ball fumes.
- Avoid perfumes and essential oils (this includes strong smelling household cleansers)
- Avoid dental drilling.
- Limit contact with cigarette smoke, especially menthol as much as possible.
- Avoid raw garlic, cooked is okay.

Chapter 5

Selected remedies for acute care of muscle injuries

While there are literally hundreds of homeopathic remedies that could be appropriate for consideration in the care of an individual with muscle injury, damage or disease, we will look at a selection of remedies that are exceptionally useful for acute care of injuries involving muscles. Remember, while homeopathic remedy selection is based solely on the symptoms of the individual and not the diagnosis, the traditional medical evaluation and care is critical. Homeopathy serves best as a true 'complementary' medicine.

Traumatic injuries that damage the muscles involve multiple factors, including damage to tissues, blood vessels, ligament, tendons cartilage and sometimes bones – where pain, inflammation and bruising are usually the overwhelming acute symptoms. Each individual will experience symptoms of type and location of pain, limitations of range of motion, weight bearing capacity, swelling or warmth, etc. in their own way. Homeopathic remedy selection is based solely on the symptoms of the individual and not the diagnosis.

Serious muscle injuries are traditionally treated by one or more of the following:

- Surgery
- Physical therapy
- Anti-inflammatory medications (e.g. NSAIDs or oral steroids)
- Stretching and strengthening exercises
- Massage therapy

Homeopathy offers a combination of effective pain relief and support for the body's own healing mechanisms to work to their maximum capacity. Using homeopathy during the healing process may greatly reduce the pharmaceutical medication required.

Acute muscle problems usually fall into one of several basic categories.

- Traumatic injury to a functioning muscle, muscle fibers or muscle connections

- Cramps
- Laceration
- Loss of control
- Pain
- Puncture
- Strain – a stretch or tear of the muscle fibers
- Combination - Almost any injury to ligaments, tendons, bones and fascia will also involve muscle injuries. When there is a combination –look to the remedy that best describes the overall situation, or the remedy that best addresses the symptoms of the individual at that moment.

Most often the acute situations involve damage to tissues, blood vessels, and sometimes bones – where pain, inflammation and bruising are the overwhelming symptoms.

 Any injuries that stem from traumatic accidents have a combination of effects on the body and mind. Any damage to the nervous and circulatory system have profound effects that are difficult for traditional medicine to treat since they affect how the individual perceives pain, as well as impacting stress systems in the body, with their related effects. This may result in drastically exaggerated pain and symptoms for the amount of physical damage that is observable, and extend to emotional symptoms that are often perceived as unrelated to the accident. Symptoms may include extreme fatigue, depression, personality changes, loss of enthusiasm, etc.

Homeopathy, being a true 'holistic' approach, takes the physical and emotional symptoms into the process of individual remedy selection. As opposed to traditional treatment protocols, in homeopathy, the individual's manifestations of symptoms are considered as more important than the 'diagnosis'. The advantage of homeopathic treatment is that there is no 'masking' of the symptoms. They are resolved or not, depending on the appropriate choice of remedy.

In using choosing homeopathic remedies, it is the 'suite' of symptoms (location and description of pain, the full range of physical symptoms, emotions, etc.) of the individual that is critical in choosing the best remedy.

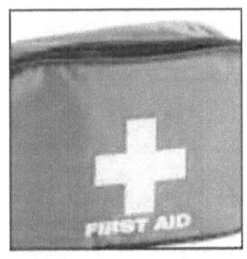 When presented with an emergency acute situation at home, in the office, on the playing field, or while hiking, there are quick – one or two word summaries that help you choose a good homeopathic 'first aid' remedy. This type of summary in no way gives the full picture of the capability or uses of the remedy – but they are handy to know, and will usually be found in a standard homeopathic first aid kit. A sample of these would be:

- Aconite for shock
- Arnica for trauma, bruising, hemorrhage
- Cantharis for burns
- Hypericum for any injury to nerve centers and tissues (including areas rich in nerves such as the fingers)
- Apis for acute synovitis
- Ledum for puncture injuries

- Silica for helping to extrude small foreign bodies (i.e. splinters or glass shards)
- Staphysagria for incised wounds

In homeopathic reference books, there are lists of 'key notes'. These are short 'snap shot' type summaries of remedies affinities – such as:

Back
- Burning between scapulae as of hot coals.
- Pain in small of back
- Lumbago
- Sciatica ameliorated by walking
- Sciatica, worse right side.

Extremities
- Numbness, also drawing and tearing in limbs, especially while at rest or at night.
- Heaviness of arms.
- Tearing in shoulder and elbow joints.
- One foot hot, the other cold.

While they are handy – they can also lead you to select too quickly, omitting consideration of a better remedy choice. Use 'key notes' only in emergencies and consult with a professional homeopath for longer term recovery and chronic conditions.

The following lists of homeopathic remedies are organized into separate sections:

- Lacerations, scrapes and burns
- Motor control
- Muscle cramps
- Muscle trauma
- Nerve injuries
- Nervousness and anxiety following trauma
- Sprains, breaks and dislocations
- Puncture wounds

They outline the common acute homeopathic remedies for situations involving muscle injuries and pain with selected keynotes.

Lacerations, scrapes and burns

Arnica Montana is the first medicine to use following all trauma and injuries. It would be used for any localized swelling in the region but is also important if there is any arterial bleeding

Calendula stops bleeding, inhibits infection and promotes healing. Not recommended for very deep cuts since it can close the outside skin before the underlying tissues have healed. Excellent for excessive pain. Usually used for cuts, burn and abrasions. Frequently used as an external application.

Cantharis is best known for use with burns and scalds and is helpful when the pain is described as 'burning'. It is useful when there is inflammatory swelling following burns.

Ignatia amara for any hysterical reaction to the bleeding or pain

Hypericum for any radiating nerve like pain from the wound site

Motor Control

Gelsemium is used when there is loss of muscular control, paralysis, trembling and weakness. Deep seated muscular pains, numb feeling as if feet were going to sleep. Chilliness up and down the back.

Muscle Cramps

Arnica Montana is the first medicine to use following all trauma and injuries. It is useful for cramps brought on by fatigue.

Calcaria Carbonica is frequently useful for cramps in legs – particularly felt in the night. Also frequently used in patients who are overweight where muscles have become lax.

Cuprum metallicum is also used for cramps of muscles especially legs, hands and calves. *Cuprum* symptoms are intense. Cramps that respond well to Cuprum frequently begin as twitches. Pains worse with touch, motion and during sleep. Improvement with pressure applied to affected area.

Magnesia phosphoricum is considered the great anti-spasmodic remedy and used for cramps of all sorts. The cramps tend to be worse on the right side, worse with cold, touch and at nighttime. There is improvement with heat applications.

Nux Vomica is useful for cramps that occur in the night. Tense cramps occur in calves and soles. Pain shoots from toes to thighs.

Muscle Trauma

Arnica Montana is the first medicine to use following all trauma and injuries. Arnica is useful for bruises, swelling, muscle soreness, sprains and strains, falls and blows, overexertion from lifting, sports, etc.

Bellis perennis is used for trauma similar to Arnica. However, Bellis is indicated for deep tissue injuries where there is intense soreness, swelling and bruising. Bellis acts on the muscular fibers of blood vessels. It is often used after Arnica.

Bryonia after Arnica when the muscle remains swollen, distended and painful and worse for motion. Intense aching that is better for pressure and cold.

Ledum Palustre is used of contusions to the eye muscles and is particularly useful for any 'black eye'. Ledum also follows Arnica well were there are bruises. It removes ecchymosis and discolorations rapidly and perfectly.

Rhus toxicondendron Excellent for pains from sprains and strains of muscle or tendon. It is also useful for bruises and general lameness. Useful when the injury might resist initial movement, it loosens up with continued movement, with pain returning with fatigue.   The injured area is better with heat and rest.

Ruta Graveolens key medicine for bruising of the lining of the bones with deep aching pain, rheumatism, tendon injuries, painful bruises and sciatic nerve pain Used for injuries and bruises to flexor tendons and the periosteum (the covering of the bone). There can also be weakness, stiffness. *Ruta* works well after *Arnica*.

Symphytum is best known to aid broken bones to mend quickly, but it also is used for facial injuries to the cheek, eye ball and area around the eyes. It has been called the 'Arnica of the eye'.

Nerve Injuries

Hypericum is excellent for injury to nerves, especially neck, back, fingers toes, nails, and brain. The characteristic pains are shooting and sharp, very sensitive to touch. Hypericum is also indicated for puncture wounds (including the site of an injection).When it is indicated following severe injuries, we will use Hypericum in conjunction with Arnica.

Ledum palustre is used for more localized nerve pain. The affected part feels cold and yet is relieved by cold application.

Ruta Graveolens is useful for painful bruises and sciatic nerve pain

Nervousness and Anxiety Following Trauma

Aconitum napellus is often useful for the anxiety and fear associated with trauma. Left untreated, this anxiety may produce or exaggerate physical symptoms.

Gelsemium is useful for athletes, performer or students who are nervous prior to a large event. (competition, audition, test, etc.) where there is a fear of appearing in public, or of failure. Gelsemium is for the trembling, diarrhea and weakness that can happen from Stage Fright. Following a trauma, Gelsemium is indicated when fear is a dominant factor. Deep seated muscular pains, numb feeling as if feet were going to sleep. Chilliness up and down the back.

Ignatia amara is similar to Aconite in that it is frequently useful for the anxiety associated with trauma which left untreated may produce or exaggerate physical symptoms. However Ignatia is usually most effective when the resulting anxiety is more related to grief or guilt.

Sprains, Strains, Breaks and Dislocations

(a *strain* involves overstretching or overuse of a muscle, ligament, or tendon; a *sprain* involves the tearing of a muscle, ligament, or tendon around a joint)

Arnica Montana is the first medicine to use following all trauma and injuries. Arnica is useful for bruises, swelling, muscle soreness, sprains and strains, falls and blows, overexertion from lifting, sports, etc.

Bellis perennis sprains with great soreness and also bruising

Bryonia is useful for sprains, dislocations, injuries to tendons and the muscle involvement that accompanies these injuries. The pains feel worse from the slightest movement, worse from being jarred or bumped and worse cold. Stiffness. Pain may be described as throbbing or stitching. Feels better keeping still and better with pressure. *Bryonia* keynotes also include irritability, desire to be alone and dry mouth. The pain is very sharp.

Calcarea carbonica – Pain deep in muscles. Swollen joints . Frequently useful for an individual who is chilly, flabby or overweight, and easily tired by exertion. Worse in cold or wet weather, physical exertion, standing, lifting, stooping and worrying. Better in dry weather, lying still, lying on painful side, and any soothing

Lycopodium – burning pain, bubbling sensation, numbness, tearing pains, twitching and jerking. Better with motion, especially restlessness better with motion, warm drinks, cool air. Worse in the afternoon and when lying on the affected side, cold food or drink, sitting erect.

Rhus toxicondendron is excellent for pains from sprains and strains of muscle or tendon. It is also useful for bruises and general lameness. Useful when the injury might resist initial movement, it loosens up with continued movement, with pain returning with fatigue.   The injured area is better with heat and rest.

Ruta graveolens is used for injuries and bruises to flexor tendons and the periosteum (the covering of the bone). *Ruta* is also indicated in the overuse of tendons (including tennis elbow), bursitis and nodular growths in the wrists. There can also be weakness, stiffness. *Ruta* works well after *Arnica*.

Symphytum is THE homeopathic remedy for broken bones. Care should be taken to insure the bone is set properly before beginning Symphytum since it encourages rapid bone healing.

Puncture Wounds

Ledum palustre is best known for its use with any
     puncture wound. From needles to nails and even
     insect stings and animal bites, it is useful in
     healing the wound and preventing tetanus as well
     as relieving the localized nerve pain of the
     puncture.

Silica will cause the body to extrude small foreign bodies
     (i.e. splinters or glass shards) that may be
     embedded after a puncture.

REMEDY SUMMARIES

Out of over 3000 homeopathic remedies currently known and documented, there are literally hundreds that might be useful in the treatment of an individual recovering from muscle injuries. We will focus on those best known for their affinity for acute situations. These few remedies are useful to alleviate suffering and speed the healing process – helping the individual pass the acute phase quickly and enter longer term healing and recovery. At that point, it is certainly recommended that a full homeopathic case may be taken and that homeopathic remedies may then be used in concert with appropriate therapies to insure the individual regains the best possible health.

A good materia medica will provide a full and complete description of the action of homeopathic remedies, but these can fill volumes. They describe the known actions and affinity of each remedy to mind and body carefully including symptoms and sensations. Depending on your needs and interests, you are encouraged to explore these fascinating books in bookstores or online by typing 'homeopathic material medica' in a search engine.

The following remedies will be presented in abbreviated form, including a very short overview, and a set of keynotes and modalities. These are drawn primarily from *Homeopathic Materia Medica* by William Boericke,

MD, originally published in 1901, with minor additions from a wide variety of additional sources.

Remedy Name	Abbreviation
Aconitum napellus	Acon
Arnica montana	Arn
Bellis perennis	Bell-p
Bryonia	Bry
Calcaria carbonica	Calc or Calc carb
Calendula	Calen
Cantharis vesicatoria	Canth
Cuprum metallicum	Cupr
Gelsemium	Gels
Hypericum perforatum	Hyper
Ignatia amara	Ign
Ledum Palustre	Led
Lycopodium	Lyc
Magnesium phosphoricum	Mag p or Mag phos
Nux vomica	Nux-v
Rhus toxicondendron	Rhus t or Rhus tox
Ruta graveolens	Ruta
Silica	Sil
Symphytum	Symph

Aconitum Napellus

A state of fear, anxiety; anguish of mind and body. Physical and mental restlessness, fright, is the most characteristic manifestation of Aconitum Napellus. Acute, sudden, and violent invasion, with fever, call for it. Does not want to be touched. Sudden and great sinking of strength. Complaints and tension caused by exposure to Dry, cold weather, draft of cold air, checked perspiration, also complaints from Very hot weather, especially gastro-intestinal disturbances, etc. First remedy in inflammation, inflammatory fevers. Serous membranes and muscular tissues affected markedly. Burning in internal parts; Tingling, coldness and numbness. Influenza. TENSION of arteries; emotional and physical mental tension explain many symptoms. When prescribing Aconite remember Aconite causes only functional disturbance, no evidence that it can produce tissue change - its action is brief and Shows no periodicity. Its sphere is in the beginning of an acute disease and not to be continued after pathological change comes. In hyperaemia, congestion not after exudation has set in.

Common names: Monkshood, Monk's Blood, Fuzi, Wolf's Bane

Selected Keynotes :

Mind
- Fearful
- Restlessness
- Panic states, these are sudden and violent
- Calm in between the panic states

Generalities
- Extreme excitability of nervous and vascular system.
- Epilepsy or convulsions

53

- Numbness, tingling agg. left side
- Intolerable pains; stinging, burning
- Protects painful part with hands
- Stiff neck muscles, especially due to draught of air

Extremities
- Acute rheumatism with sensation of enlargement of the part
- Affections often associated with numbness and tingling
- Paralysis, from exposure to cold, dry wind, from fright; hysterical.
- Paralysis on waking, unable to move, produces anxiety or even panic.

Modalities:

BETTER in open air.

WORSE in warm room, in evening and night; worse lying on affected side, with motion, from music, from tobacco-smoke, dry cold winds and dry cold weather.

Arnica montana

Produces conditions upon the system quite similar to those resulting from injuries, falls, blows, contusions. Tinnitus aurium. Putrid phenomena. Septic conditions; prophylactic of pus infection. Apoplexy, red, full face. It is especially suited to cases when any injury, however remote, seems to have caused the present trouble. After traumatic injuries, overuse of any organ, strains. Arnica Montana is disposed to cerebral congestion. Acts best in plethoric, feebly in debilitated with impoverished blood, cardiac dropsy with dyspnoea. A muscular tonic. Traumatism of grief, remorse or sudden realization of financial loss. Limbs and body ache as if beaten; joints as if sprained. Bed feels too hard. Marked effect on the blood. Affects the venous system inducing stasis. Ecchymosis and Hemorrhages. Relaxed blood vessels, black and blue spots. Tendency to hemorrhage and low-fever states. Tendency to tissue degeneration, septic conditions abscesses that do not mature. Sore, lame, bruised feeling. Neuralgias originating in disturbances of pneumogastric. Rheumatism of muscular and tendinous tissue, especially of back and shoulders. Aversion to tobacco. Influenza. Thrombosis. Haematocele.

Source: Leopard's Bane

Selected Keynotes

Ailments from injury and trauma

Mind
- Aversion to being touched
- Says he is well, thinks he is well, sends doctor away
- Sensitive to pain.
- Defensive: Irritable, obstinate when approached.
- Restless because of soreness.

Generalities
- Complaints ever since a TRAUMA
- AILMENTS FROM INJURIES WITH A BLUNT
  INSTRUMENT, OVEREXERTION and surgery
- SORE, BRUISED PAINS.
- Tendency to hemorrhages, externally and internally.
- Any bruised muscle
- Cramps which occur when fatigued
- Muscle stiffness after exertion and strenuous games

Back
- Muscles of neck weak, head falls backwards or to
  the side
- Cervical vertebrae tender
- General soreness of the back
-

Extremities
- Arthritis, rheumatism
- Joint pains
- SPRAINS. BRUISES
- Limbs ache as if beaten
- Knee joint suddenly bends when standing
- Feet numb

Modalities:

BETTER, lying down, or with head low.

WORSE, least touch; motion; rest; jarring; cold
weather, damp cold.

An interesting side note concerning Arnica – the 'picture' of
arnica is the individual who, while holding a broken arm, will
say 'I'm alright' or 'no problem' and brush help aside. They do
not melt into a sobbing puddle and want to be held and
stroked.

Bellis perennis

It acts upon the muscular fibers of the blood-vessels. Much muscular soreness. Lameness, as if sprained. Venous congestion, due to mechanical causes. First remedy in injuries to the deeper tissues, after major surgical work. Results of injuries to nerves with intense soreness and intolerance of cold bathing. After gout, debility of limbs. Traumatism of the pelvic organs, auto-traumatism, expresses the condition calling for this remedy. Excellent remedy for sprains and bruises. Complaints due to cold food or drink when the body is heated, and in affections due to cold wind. Externally, in naevi. Acne. Boils all over. sore, bruised feeling in the pelvic region. Eructations, stasis, swelling, come within the range of this remedy. Rheumatic symptoms. Does not vitiate the secretions. "It is a princely remedy for old laborers, especially gardeners." (Burnett.)

Common name: Common Daisy

Selected Keynotes
Sprains, bruises, injuries, bone fractures

Mind
- Obstinate. Fixed ideas.
- Confusion of surroundings.

Generalities
- Sprains, bruises, injuries, bone fractures
- Injuries to deeper tissues after major operation
- Ailments after getting cold when heated.
- Overworked old laborers, gardeners, commercial travelers,
- railway spine.

Extremities

- Healing of cuts and wounds
- Deep tissue pain
- Sprains with great soreness
- Wrist feels contracted as from elastic band around
  joint
- Pain down anterior of thighs
- Joints sore, muscular soreness
- Injuries to nerves, with intense soreness

Modalities

WORSE left side; overexertion, hot bath and warmth
of bed; before storms; cold bathing; cold wind.

Bryonia

Acts on all serous membranes and the viscera they contain. Aching in every muscle. The general character of the pain here produced is a stitching, tearing; worse by motion, better rest. These characteristic stitching pains, greatly aggravated by any motion, are found everywhere, but especially in the chest; worse pressure. Mucous membranes are all dry. The Bryonia Alba patient is irritable; has vertigo from raising the head, pressive headache; dry, parched lips, mouth; excessive thirst, bitter taste, sensitive epigastrium, and feeling of a stone in the stomach; stools large, dry, hard; dry cough; rheumatic pains and swellings; dropsical effusions into synovial and serous membranes. Bryonia Alba affects especially the constitution of a robust, firm fiber and dark complexion, with tendency to leanness and irritability. It prefers the right side, the evening, and open air, warm weather after cold days, to manifest its action most markedly. Physical weakness, all-pervading apathy. Complaints apt to develop slowly.

Common names: White Bryony, Wild Hops, Cucurbitaceae, English Mandrake, Tetterbury

Selected Keynotes

Mind
- Aversion to being disturbed.
- Irritable, wants to be left alone. Reserved.
- Thoughts, talks about business. Determined.
- Despair of recovery. Seem to hold others responsible for their suffering.
- Dullness morning on waking,

Generalities
- Dryness of mucous membranes
- Often slow onset of acute diseases.

- Injuries, sprains, etc.
- Stitching pains.
- Slow onset
- Great thirst for large quantities
- Also possible: dry mouth and thirst less

Back
- Painful stiffness in nape of neck
- Stitches and stiffness in small of back

Extremities
- Rheumatism of joints and muscles, arthritis, gout
- Inflammation of joints with swelling and stiffness.
- Pain that feels like bone in hurting

Modalities

BETTER, lying on Painful side, pressure, rest, cold applications.

WORSE, warmth, any motion, morning, eating, heat, hot weather, jarring, exertion, touch. Cannot sit up-gets faint and sick.

Calcaria carbonica

This great Hahnemannian anti-psoric is a constitutional remedy Par Excellence. Its chief action is centered in the vegetative sphere, impaired nutrition being the keynote of its action, the glands, skin and bone, being instrumental in the changes wrought. Increased local and general perspiration, swelling of glands, scrofulous and rachitic conditions generally offer numerous opportunities for the exhibition of Calcarea Carbonica. Incipient phthisis. It covers the tickling cough, fleeting chest pains, nausea, acidity and dislike of fat. Gets out of breath easily. A jaded state, mental or physical, due to overwork. abscesses in deep muscles; polypi and exostoses. Pituitary and thyroid disfunction. Is a definite stimulant to the periosteum. Easy relapses, interrupted convalescence. Persons of scrofulous type, who take cold easily, with increased mucous secretions, children who grow fat, are large-bellied, with large head, pale skin, chalky look, the so-called leucophlegmatic temperament; affections caused by working in water. Great sensitiveness to cold; partial sweats.

Source: Oyster shell

Common name: Calcium carbonate

Selected Keynotes

Mind
- Hard working, overworking, capable, conscientious, over-responsible, take on too much.
- Practical, down to earth.
- Obstinate.
- Many small fears. Worry about small things
- Anxiety about health, future.
- Despair of recovery
- Melancholy

Generalities
- Pillow wet from perspiration neck.
- Chilly
- Sour odor of discharges, of body.
- Bones: Weak, disturbed development, caries.

Back
- Weakness of back.
- Curvature of spine.
- Lumbago, sciatica.
- Swelling of cervical glands.

Extremities
- Weakness- Sprains easily.
- Arthritis, rheumatism
- Cold, clammy hands and feet.
- Nails brittle
- Cramps in calves at night in bed.

Modalities

BETTER, dry climate and weather; lying on painful side. Heat of the sun.

WORSE, from exertion, mental or physical; ascending; cold in every form; water, washing, moist air, wet weather; during full moon; standing.

Calendula Officinalis

A most remarkable healing agent, applied locally. Useful for open wounds, parts that will not heal, ulcers, etc. Promotes healthy granulations and rapid healing by first intention. Hemostatic after tooth extraction. Deafness. Catarrhal conditions. Neuroma. Constitutional tendency to erysipelas. Pain is excessive and out of all proportion to injury. Great disposition to take cold, especially in damp weather. Paralysis after apoplexy. Cancer, as an intercurrent remedy. Has remarkable power to produce local exudation and helps to make acrid discharge healthy and free. Cold hands.

Common names: Marigold

Selected Keynotes

Healing of wounds

Mind
- Extremely nervous. Sensitive to noise
- Easily frightened
- Restlessness with fear of impending misfortune
- Irritable

Generalities
- Wounds – raw, inflamed, red, stinging
- Open, cut, lacerated, ragged, suppurated
- Promotes healthy granulations and rapid healing
- Exhaustion from loss of blood or excessive pain
- Pain is excessive and out of all proportion to injury.
- Prevents lymphangitis, gangrene, sepsis.
- Injuries of muscles, tendons.

Extremities
- Weakness, particularly in lower limbs

Modalities

WORSE, in damp, Heavy, cloudy weather.

Cantharis vesicatoria

This powerful drug produces a furious disturbance in the animal economy, attacking the urinary and sexual organs especially, perverting their function, and setting up violent inflammations, and causing a frenzied delirium, simulating hydrophobia symptoms. Puerperal convulsions. Produces most violent inflammation of the whole gastro-intestinal canal, especially lower bowel. Oversensitiveness of all parts. Irritation. Raw, burning pains. Hemorrhages. Intolerable, constant urging to urinate is most characteristic.

Common names: Spanish fly

Selected Keynotes

Burns and Scalds

Mind
- Restlessness
- Contradictory mood

Generalities
- Violent, burning pains, also cutting and biting pain
- Oversensitiveness of all parts
- Action is rapid and intense
- inflammations are violently acute

Extremities
- burning soles of feet at night
- knees totter when ascending steps

Modalities

WORSE, coffee, drinking anything cold, touch

BETTER, warmth, rest, rubbing

Cuprum metallicum

Spasmodic affections, cramps, convulsions, beginning in fingers and toes, violent, contractive, and intermitting pain, are some of the more marked expressions of the action of Cuprum Metallicum; and its curative range therefore includes tonic and clonic spasms, convulsions, and epileptic attacks. Chorea brought on by fright. Nausea greater than in any other remedy. In epilepsy, aura begins at knees, ascends to hypogastrium; then unconsciousness, foaming, and falling. Symptoms disposed to appear periodically and in groups. Complaints begin in left side. The pains are increased by movement and touch.

Common names: Metallic copper

Selected Keynotes
Cramps on all levels

Mind
- Nervous
- Uneasy
- Loquacious – then melancholy

Generalities
- Symptoms appear in groups
- Most often complaints begin on left side
- Continued weakness

Extremities
- Affects the nerves of the cerebrospinal axis and muscles – causing spasmodic effects
- Convulsions and cramps are violent
- Twitchings. Chorea
- Clenching thumbs and fingers
- Convulsions may be tonic or clonic
- Paralysis of isolated muscles

- Cramps in palms, calves and soles
- Joints contracted
- Jerking in hands and feet

Modalities

WORSE, touch and motion, hot weather, strong emotions (i.e. anger or fright)

BETTER, cold drinks

Gelsemium

sempervirens

Centers its action upon the nervous system, causing various degree of Motor paralysis. General prostration. Dizziness, drowsiness, dullness, and trembling. Slow pulse, tired feeling, mental apathy. Paralysis of various groups of muscles about the eyes, throat, chest, larynx, sphincter, extremities, etc. Post-diphtheritic paralysis. Muscular weakness. Complete relaxation and prostration. Lack of muscular coordination. General depression from heat of sun. Sensitive to a falling barometer; cold and dampness brings on many complaints. Sluggish circulation. Nervous affections of cigar makers. Influenza. Measles. Pellagra.

Common name: Yellow jasmine, evening trumpetflower, woodbine

Selected Keynotes

Mind
- Feeling of weakness, of not being able to cope with daily life, responsibilities, work.
- Timidity.
- Desire to be quiet, to be left alone. Avoid people and distress of life.
- Fear of falling, crowds.
- Mental weakness. Dullness. Forgetful.

Generalities
- Weakness, paresis, paralysis
- Ailments from anticipation, fright, excitement, bad news
- Trembling from weakness, fear, anticipation .
- Flushes of heat alternating with chills, -

Back
- Chilliness up and down the back
- Weakness.
- Pain and stiffness cervical region extending to head.
- Pain under left scapula.

Extremities
- Loss of power of muscular control.
- Paralysis. Contractions.
- Trembling and weakness
- Heavy feeling, especially lower limbs.

Modalities

BETTER, bending forward, by profuse urination, open air, continued motion, stimulants.

WORSE, damp weather, fog, before a thunderstorm, emotion, or excitement, Bad news, tobacco-smoking, when thinking of his ailments; at 10 A.m.

Hypericum perforatum

The great remedy for injuries to nerves, especially of fingers, toes and nails. Crushed fingers, especially tips. Excessive painfulness is a guiding symptom to its use. Prevents lockjaw. Punctured wounds. Relieves pain after operations. Spasms after every injury. Has an important action on the rectum; hemorrhoids. Coccydynia. Spasmodic asthmatic attacks with changes of weather or before storms, better by copious expectoration. Injured nerves from bites of animals. Tetanus. Neuritis, tingling, burning and numbness. Constant drowsiness.

Common name: St. John's Wort

Selected Keynotes

> **Injuries to nerves and spine.**

Mind
- Symptoms after injuries include: depression, dullness, forgetfulness, hysteria
- Feels as if lifted in the air.
- Fear of falling from height, downward motion

Generalities
- INJURIES TO PARTS RICH IN NERVES
- Pain: shooting upward along nerve
- Convulsions after injury, esp. to spine, head
- Ailments of shock, fright
- Phantom pains

Back
- Injuries to spine and coccyx

- Pain shoots up the spine and down limbs.
- Bruised coccyx after labor.

Extremities
- Neuritis.
- Injuries to finger tips.
- Rheumatic pains.

Modalities

BETTER, bending head backward.

WORSE, in cold; dampness, in foggy weather; before thunderstorm; in close room; least exposure; touch.

Ignatia amara

Produces a marked hyperaesthenia of all the senses, and a tendency to clonic spasms. Mentally, The emotional element is uppermost, and co-ordination of function is interfered with. Hence, it is one of the chief remedies for hysteria. It is especially adapted to the nervous temperament. Rapid change of mental and physical condition, opposite to each other. Great contradictions. Alert, nervous, apprehensive, rigid, trembling patients who suffer acutely in mind or body, at the same time made worse by drinking coffee. The superficial and erratic character of its symptoms is most characteristic. Effects of grief and worry. Cannot bear tobacco. Pain is small, circumscribed spots . The plague. Hiccough and hysterical vomiting.

Source: Strychnos ignatii

Common name: St. Ignatius Bean

Selected keynotes

Mind
- Emotional sensitivity
- Ailments from grief, disappointment.
- Brooding silent grief
- Changeable moods. Unpredictable, hysterical
- Anxiety about health.

Generalities
- Paradoxical, contradictory symptoms
- Hysterical tendency and symptoms
- Numbness. Cramps. Spasms. Tics. Chorea.
- Sudden momentary paralysis.

Back
- Spasms.

Extremities
- Jerks, twitchings, chorea.
- Hysterical numbness, tingling, paralysis
- Cramps and spasms

Modalities

BETTER, while eating, change of position

WORSE, in the morning, open air, after meals, coffee, smoking, liquids, external warmth.

Ledum Palustre

Affects especially the rheumatic diathesis, going through all the changes, from functional pain to altered secretions and deposits of solid, earthy matter in the tissues. The Ledum Palustre rheumatism begins in feet, and travels upward. It affects also the skin, producing an eruption like Poison-oak, and is antidotal thereto, as well as to stings of insects. There is a general lack of animal heat, and yet heat of bed is intolerable. For punctured wounds, produced by sharp-pointed instruments or bites, particularly if the Wounded parts are cold, this is the remedy. Tetanus with twitching of muscles near wound.

Common names: Marsh Tea, Labrador Tea

Selected Keynotes

Rheumatic complaints. Punctured wounds, stings and bites.

Mind
- Irritable. Dissatisfied
- Restlessness, must constantly move

Generalities
- Chilly but aggravated by heat
- Pain ameliorated by cold
- Coldness of affected part
- Complaints go from below upward
- Ailments from puncture wounds
- Prevents tetanus
- Tetanus with twitching of muscles near wound

Extremities
- Rheumatism and gout, mainly of lower limbs
- Affections travel upward from feet

- Swelling of affected joints
- Cross-wise pains (example, left shoulder and right hip)
- Itching of feet and ankles
- Elephantiasis of lower extremities
- Easy spraining of ankles
- Muscles involved in stab wounds and hypodermic injections
- Muscle stiffness due to cold and damp
- Old yellow bruises
- Bites, stings, punctured wounds

Modalities:

BETTER, from cold applications, putting feet in cold water.

WORSE, at night, and from heat of bed.

Lycopodium clavatum

In nearly all cases where Lycopodium is the remedy, some evidence of urinary or digestive disturbance will be found. Lycopodium is adapted more especially to ailments gradually developing, functional power weakening, with failures of the digestive powers, where the function of the liver is seriously disturbed. Atony. Malnutrition. Mild temperaments of lymphatic constitution, with catarrhal tendencies; older persons, where the skin shows yellowish spots, earthy complexion, uric acid diathesis, etc; also precocious, weakly children. Symptoms characteristically run from right to left, acts especially on right *side of body, and are worse from about 4 to 8 pm. In kidney* affections, red sand in urine, *backache, in renal region; worse before urination. Intolerant of cold drinks;* craves everything warm. *Best adapted to persons intellectually keen, but of weak, muscular power. Deep-seated, progressive, chronic diseases. Carcinoma. Emaciation. Debility in morning. Marked regulating influence upon the glandular (sebaceous) secretions. Pre-senility. Ascites, in liver disease. Lycop patient is thin, withered, full of gas and dry. Lacks vital heat; has poor circulation, cold extremities. Pains come and go suddenly. Sensitive to noise and odors.*

Common Name: Club moss

Selected keynotes

Mind
- Often intellectual people like schoolteachers, lawyers.
- Timid
- Presents themselves different than they feel Inside
- Extremely sensitive

Back
- Burning between scapulae as of hot coals.
- Pain in small of back
- Torticollis
- Lumbago
- Sciatica ameliorated by walking
- Sciatica, worse right side.

Extremities
- Numbness, also drawing and tearing in limbs, especially while at rest or at night.
- Heaviness of arms.
- Tearing in shoulder and elbow joints.
- One foot hot, the other cold.
- Chronic gout, with chalky deposits in joints.
- Profuse sweat of the feet.
- Pain in heel on treading as from a pebble.
- Painful callosities on soles;
- Toes and fingers contracted.
- Cannot lie on painful side.
- Hands and feet numb.
- Right foot hot, left cold.
- Cramps in calves and toes at night in bed.
- Limbs go to sleep.
- Twitching and jerking.

Modalities

BETTER, by motion, after midnight, from warm food and drink, on getting cold, from being uncovered.

WORSE, right side, from right to left, from above downward, 4 to 8 pm; from heat or warm room, hot air, bed. Warm applications, except throat and stomach which are better from warm drinks.

Magnesium phosphoricum

The great anti-spasmodic remedy. Cramping of muscles with radiating pains. Neuralgic pains relieved by warmth. Especially suited to tired, languid, exhausted subjects. Indisposition for mental exertion. Goitre.

Selected Keynotes

Mind
- Oversensitive.
- Lamenting all the time about pain.
- Irritable.
- Capricious.
- Indisposition to mental exertion.

Generalities
- Neuralgic pains
- General muscular weakness

Neck and back
- Stiffness of neck and back
- Cramps
- One vertebra seems absent

Extremities
- Cramps from prolonged exertion.
- Sciatica,
- Weakness upper limbs, trembling. Parkinson.
- Cramping pains
- Cramps in calves
- Spasmodic pain
- Acute stabbing pain
- Boring pain

Modalities

BETTER for warmth, bending double, pressure, friction.

WORSE on right side, COLD, touch, night.

Nux vomica

Is the greatest of polychrests, because the bulk of its symptoms correspond in similarity with those of the commonest and most frequent of diseases. It is frequently the first remedy, indicated after much dosing, establishing a sort of equilibrium of forces and counteracting chronic effects.

Nux is pre-eminently the remedy for many of the conditions incident to modern life. An Irritable, nervous system, hypersensitive and over impressionable, which Nux will do much to soothe and calm. Especially adapted to digestive disturbances, portal congestion, and hypochondriacal states depending thereon. Convulsions, with consciousness; worse, touch, moving. Zealous fiery temperament. Nux patients are easily chilled, avoid open air, etc. Nux always seems to be out of tune; inharmonious spasmodic   *action.*

Source: Strychnos nux-vomica

Common names: Poison nut

Selected Keynotes:

Mind
- Responsible. Efficient
- Fastidious.
- Over-stressed
- Oversensitive.
- Irritability
- Impatient

Generalities
- Chilly
- Sensitive to all impressions: odors, light, noise, smells, etc.
- Lightning-like pains
- Neuralgia
- Reflexes increased

Back
- Pain, must sit up to turn over in bed
- Opisthotonos
- Cervico-brachial neuralgia
- painfully stiff neck
- Pains travel down shoulder
- Lumbar ache, as if breaking
- Crawling sensation along spine
- Acute lumbago
- Sacral region feels lame

Extremities
- Muscle spasms, cramps and contractures
- Twitchings. Jerkings.
- Cramps that begin in the night
- Cracking in knee joints during motion
- Paralysis of lower limbs from over exertion
- Stiff legs
- Feet feel clubby and raw
- Drags feet while walking
- Legs tremble
- Unsteady gait

Modalities

BETTER from a nap, if allowed to finish it; in evening, while at rest, in damp, wet weather, strong pressure, warmth

WORSE in the morning, mental exertion, after eating, touch, spices, stimulants, narcotics, dry weather, cold.

Rhus toxicondendron

The effects on the skin, rheumatic pains, mucous membrane affections, and a typhoid type of fever, make this remedy frequently indicated. Rhus Toxicodendron affects fibrous tissue markedly - joints, tendons, sheaths - aponeurosis, etc., producing pains and stiffness. Post-operative complications. Tearing asunder pains. Motion always "limbers up" the Rhus patient, and hence he feels better for a time from a change of position. Ailments from strains, overlifting, getting wet while perspiring. Septic conditions. Cellulitis and infections, carbuncles in early stages . Rheumatism in the cold season. Septicaemia.

Common name: Poison ivy

Selected Keynotes

Mind
- Busy, restless
- Timid and mild, yet lively.
- Later on irritable, impatient.
- Depressed, morose.
- Becomes rigid and stiff as with the muscles and joints.
- Behavior disorders with restlessness, irritability.

Generalities
- Sudden periodic chills esp. at night.
- Parkinson's disease.

Back
- Pain and stiffness
- Stiffness of the neck,

Extremities
- Arthritis, rheumatism with painful stiffness
- Pain Left shoulder , inner side left scapula.

- Restless extremities. Restless legs in bed at
  night.
- Sprains. Overexertion.
- Bursitis. Tendonitis.
- Chorea. Twitching. Tics.
 - Joint pains and fevers

Overstretching or overexertion leaves muscles
inflamed, stiff and in need of gentle motion and
warmth

Sore muscles, bruised feeling, strained back muscles.
Needs to stretch and move. Massage neck constantly.
Muscles bound up in neck.

Modalities

BETTER, warm, dry weather, motion, walking, change
of position, rubbing/massage, warm applications, from
stretching out limbs.

WORSE, overexertion, during sleep, cold, wet rainy
weather and after rain, foggy weather, thunderstorms,
at night, during rest, draft, when lying on back or right
side.

Ruta graveolens

Acts upon the periosteum and cartilages, eyes and uterus. Complaints from straining Flexor tendons especially. Tendency to the formation of deposits in the periosteum, tendons, and about joints, especially wrist. Over strain of ocular muscles. All parts of the body are painful, As if bruised. Sprains (after Arnica). Lameness after sprains. Jaundice. Feeling of intense lassitude, weakness and despair. Injured "bruised" bones.

Common name: Rue

Selected Keynotes

Stiffness. Inflexibility (mental/emotional and physical)

Mind
- The mind loses flexibility.
- Fogginess, slowness.
- Irritability, discontented, quarrelsome
- Difficult thinking and communicating.
- Fears and anxiety

Generalities
- Stiffness – loss of elasticity of muscles and tendons.
- Weakness, easily tired
- Bruised pains

Back
- Pain is relieved by lying on back, and pressure
- Drawing pain and stiffness in neck.
- Stiffness and pain making breathing difficult

Extremities
- Rheumatic pains
- Stiffness with bruised pains and weakness
- Stiffness, compelled to move limbs or walk, but aggravated by the motion.
- Weakness of lower extremities.
- Trauma or sprain of the ligaments
- Hard nodules in tendons and periost from overuse and injury
- Inflammation and contraction of tendons
- Pain in thighs when stretching legs.

Key medicine for bruising of the lining of the bones with deep aching pain, rheumatism, tendon injuries, painful bruises and sciatic nerve pain. Bruises to bone surfaces – especially thinly covered bones such as elbows, wrists, spine and shins

Modalities

BETTER - gentle movement, indoors, warmth, massage

WORSE - from cold, wet weather, lying down, sitting, strenuous exercise

Silica

Imperfect assimilation and consequent defective nutrition. It goes further and produces neurasthenic states in consequence, and increased susceptibility to nervous stimuli and exaggerated reflexes. Diseases of bones, caries and necrosis. Organic changes; it is deep and slow in action. Periodical states; abscesses, quinsy, headaches, spasms, epilepsy, feeling of coldness before an attack. Ill effects of vaccination. Suppurative processes. It is related to all pustulous burrowings. Ripens abscesses since it promotes suppuration. Silica patient is cold, chilly, hugs the fire, wants plenty warm clothing, hates drafts, hands and feet cold, worse in winter. Lack of vital heat. Prostration of mind and body. Great sensitiveness to taking cold. Intolerance of alcoholic stimulants.

Source: flint

Selected Keynotes

Mind
- Want of self-confidence.
- Complaints from long mental exertion.

Generalities
- Chilly
- Stitching pains, splinter-like
- Frequent colds and infections with enlarged glands, that may suppurate.
- Weakness, lack of energy.
- Sensitivity
- Diseases of bone.

Back
- Diseases of bone and spine - Scoliosis. Curvatures. Caries.
- Weakness.

- Lumbago.
- Sciatica.
- Coldness. Sensitive to draft.
- Ailments from injury to the spine
- Irritation of the spine

Extremities
- Coldness.
- Nails break easily
- Pain in ankles and wrists
- Bunions
- Profuse, offensive, corrosive foot sweat.
- Arthritis

Pain arising from neglected injuries

Modalities

BETTER for warmth, wrapping up head, summer; warm food and drinks, in wet or humid weather.

WORSE in morning, from washing, uncovering, lying down, damp, lying on left side, cold, weather changes, draft, mental or physical exertion. In acute situations, heat may aggravate.

Symphytum Officinale

The root contains a crystalline solid, that stimulates the growth of epithelium on ulcerated surfaces. It may be administered internally in the treatment of gastric and duodenal ulcers. Also in gastralgia, and externally in pruritus ani. Injuries to sinews, tendons and the periosteum. Acts on joints generally. Neuralgia of knee. Of great use in wounds penetrating to perineum and bones, And in non-union of fractures; irritable stump after amputation, irritable bone at point of fracture. Psoas abscess. Pricking pain and soreness of periosteum.

Common names: Comfrey and Knit Bone

Selected Keynotes

Injuries of bones

Generalities
- Fractures – improves reunion of bones
- Pains in bones or periost after injuries.
- Injuries of periost, bones.
- Phantom pains

Back
- Backache from violent exertion.
- Caries of vertebrae.

Extremities
- Neuralgia of the knee after injury.
- Painful weakness of joints.
- Particularly helpful after broken bone has been set

Modalities

BETTER for warmth

WORSE for touch or pressure

Notes

Notes

Notes

Notes

www.ingramcontent.com/pod-product-compliance
Lightning Source LLC
Chambersburg PA
CBHW031248280526
45784CB00004B/1764